AMERICAN MUSEUM OF NATURAL HISTORY

ROSE CENTER FOR EARTH AND SPACE

A MUSEUM FOR THE TWENTY-FIRST CENTURY

AMERICAN MUSEUM OF NATURAL HISTORY

ROSE CENTER FOR EARTH AND SPACE

A MUSEUM FOR THE TWENTY-FIRST CENTURY

HAYDEN PLANETARIUM

CULLMAN HALL OF THE UNIVERSE

GOTTESMAN HALL OF PLANET EARTH

THE AMERICAN MUSEUM OF NATURAL HISTORY IN ASSOCIATION WITH HARRY N. ABRAMS, INC., PUBLISHERS

ACKNOWLEDGMENTS

We would especially like to thank Frank Summers of the Physical Sciences Division, Astrophysics/Hayden Planetarium, and Edmond A. Mathez, James Webster, and Margaret Carruthers of the Physical Sciences Division, Earth and Planetary Sciences, for their careful review of this book and help in producing it.

For the American Museum of Natural History:
  Editorial Director, AMNH Special Publishing: Maron L. Waxman
For Harry N. Abrams, Inc.:
  Project Manager: Eric Himmel
  Editor: Sharon AvRutick
  Designer: Dana Sloan

Library of Congress Card Number: 00–111542
ISBN 0–8109–2969–4

Printed and bound in Hong Kong

Harry N. Abrams, Inc.
100 Fifth Avenue
New York, N.Y. 10011
www.abramsbooks.com

Pages 2–3: Messier 100, a grand-design spiral galaxy
Opposite: The Hayden sphere

CONTENTS

FOREWORD

You are about to embark on an amazing journey. In this book, you will explore one of the most spectacular, ambitious, and thrilling projects in the history of the American Museum of Natural History—the new Frederick Phineas and Sandra Priest Rose Center for Earth and Space. This extraordinary facility embodies a new vision of what it means to be a museum in the twenty-first century and reveals the expanding role of institutions like this one in educating the public about science and the natural world, and humanity's place in it.

How did this project come about? The original Hayden Planetarium, built in 1935, was a state-of-the-art facility when it was constructed. It reflected the most advanced knowledge of our cosmos in an era that pre-dated moon launches and deep space probes. When the original Hayden Planetarium was built in 1935, not only had we not seen quasars, pulsars, and black holes, we did not even know that such phenomena could exist. Given the powerful advances in astrophysics that had taken place over the next years, it was clear by the early 1990s that the Hayden could no longer remain as a leader in cosmic information without significant updating and physical renovation.

In early 1994, we here at the Museum began to examine how the Hayden might be renovated to best reflect today's accelerated pace of scientific discovery and, equally important, how we as an institution might contribute to this exciting and fruitful era of discovery. Our goal of bringing the cosmos and the science of astrophysics to the public required not only a major reconfiguration of the physical building, but also a new intellectual framework for the exhibitions. We needed to build a strong and dedicated Department of Astrophysics to provide the rigorous scientific foundation for what promised to be thrilling but extremely challenging exhibition material. Today our Department of Astrophysics is composed of highly distinguished astrophysicists with expertise ranging from the birth of stars to the modeling of cosmic collisions. The input of our curators and scientists in the Department of Astrophysics has been invaluable not only in the creation of the Rose Center, but in its ongoing vitality and relevance in this rapidly developing branch of scientific study.

Once we began the planning process to update the Hayden, however, we were further challenged by the constraints of the original physical plant. Its design left little room for the kinds of sweeping exhibitions, interactive components, and cutting-edge technologies we needed to best present the grandeur and sweeping nature of the ever-expanding study of cosmology. James Stewart Polshek's brilliant conception of a completely reconstructed Hayden with a full dome—a sphere floating in a glass box—addressed those needs and offered a physical manifestation of our new vision for the planetarium. Thus, the Frederick Phineas and Sandra Priest Rose Center for Earth and Space was born. Thanks to the support of an extraordinarily dedicated team of trustees, government officials, local community groups, interested individuals, and Museum staff, this vision was realized in early 2000.

The Rose Center, which opened to the public on February 19, 2000, consists of a completely rebuilt and reimagined Hayden Planetarium, as well as the Gottesman Hall of Planet Earth, which opened in June 1999, and the new Cullman Hall of the Universe. This new facility adds nearly 350,000 square feet to the existing Museum—an addition that itself, in many cities, would be the largest museum. The Rose Center addresses areas of science—astrophysics, astronomy, and geology— that are experiencing a true "golden age," filled with myriad new discoveries and theories about the Big Bang—the origin of the universe—and the structure of our planet and the universe. To put this into context, the new Rose Center aims to make astrophysics, astronomy, and geology—exceedingly complex, abstract areas of science—accessible and comprehensible to the public, and to bring the frontiers of outer space and discovery to the people. By taking our visitors on a journey that reveals the inner workings of our planet and the grandeur of the universe all around us, we hope to illuminate the magnitude, the majesty, and the mystery of the cosmos, and, ultimately, humanity's place in it.

With the benefit of the one-of-a-kind Hayden Edition Zeiss Projector, as well as all-dome digital technology and a revolutionary new digital map of our galaxy, visitors to the Hayden Planetarium now can travel to the stars and beyond. The new Hayden Planetarium also is capable of receiving live feeds from NASA and other sources, allowing us to share the latest events and discoveries in outer space as they occur. In turn, we are able to transmit these images, along with our scientists' interpretations of them, from the Rose Center to classrooms, homes, and community centers across the city and the nation.

Together with our beloved existing exhibition halls, the Rose Center enables the Museum to take visitors on a grand and all-encompassing journey that tells a coherent, comprehensive story of life, from the outer reaches of the universe to the inner

core of our own fragile planet, through the extraordinary diversity of life and human culture on Earth. Put simply, the Rose Center aims to perpetuate and extend dramatically the role of the museum in communicating science to the public and advancing science literacy throughout the nation. Our extensive collections have long had the capacity to provide retrospective analysis. Now, our scientific research and cutting-edge technological capacities produce up-to-the-minute observations, including dramatic live images. Truly, this is the launch of the Museum for the new millennium. We look forward to welcoming you to the Rose Center for Earth and Space.

Ellen V. Futter, president,
American Museum of Natural History

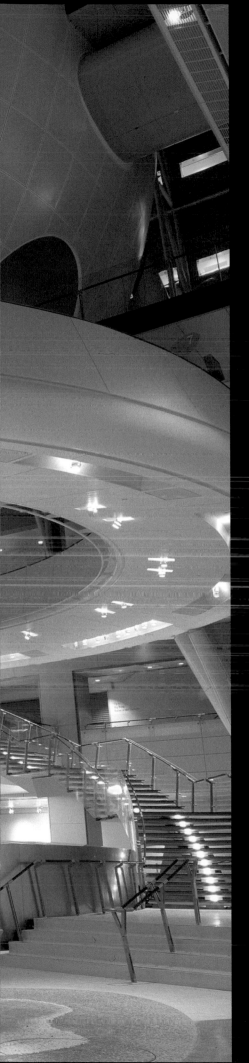

HALL OF THE UNIVERSE

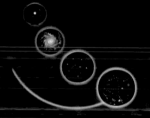

PASSPORT TO THE
UNIVERSE

Our Cosmic Address: American Museum of Natural History, Manhattan Island, North America, Earth, Solar System, Orion Spiral Arm, Milky Way Galaxy, Local Group of Galaxies, Virgo Super-cluster of Galaxies, Observable Universe, Entire Universe

The universe began with an explosion of space itself—the Big Bang. Starting from extremely high density and temperature, space expanded. As it cooled, the simplest elements formed. Gravity gradually drew matter together to form the first stars and the first galaxies. Galaxies collected into groups, clusters, and superclusters. Some stars died in supernova explosions; their chemical remnants seeded new genera-tions of stars and enabled the formation of rocky planets. On at least one such planet, life evolved to consciousness. And it wonders, "Where did I come from?"

The stunning discoveries of modern astrophysics and space exploration have opened new windows on this very human question. At the same time, these discoveries have revealed to us the basic prin-ciples at work in the universe, principles that operate in our everyday lives and in the farthest reaches of the cosmos. The Dorothy and Lewis B. Cullman Hall of the Universe explores the wonders of astron-omy and our ways of understanding them.

Introducing the space show at the Hayden Planetarium...

We are living in the golden age of astronomy. We are completing the spacecraft reconnaissance of our solar system. We are using many kinds of light to observe the life cycles of stars. We are mapping the grand structure of the universe.

The Zeiss Mark IX at the Rose Center is the most advanced star projector in the world. But to explore the universe in three dimensions, we need a powerful computer loaded with real astronomical data from the great observatories on Earth and in space. What is presented in the Rose Center space show is not an artist's fantasy, but a three-dimensional map of the real universe, carefully calculated and drawn from the best astronomical observations and data.

**The Zeiss Mark IX Hayden Edition
projector in the Space Theater of
the Hayden Planetarium.**

The entrance to the Big Bang Theater. The Big Bang is the scientific theory for the origin of the cosmos. It describes the beginning of the universe as an explosion of space, which has expanded and cooled ever since. If we rewind the expansion of space and go backward in time, galaxies crowd closer and closer together in the past. Eventually, all matter and energy is compressed at extremely high density and temperature. Today's observable universe fits within a volume smaller than a grain of sand. That super-dense medium exploded in the Big Bang, launching the expansion of the universe.

Opposite: The Harriet and Robert Heilbrunn Cosmic Pathway, some 360 feet long, is the 13-billion-year record of cosmic history, beginning with the Big Bang. On the Cosmic Pathway, 1 foot equals 45 million years, and 1 meter equals 147 million years.

Top: Around 4 billion years ago, the planets in the early Solar System were continually bombarded by comets and asteroids. Bottom: The radio waves from the galaxy shown on this station of the Cosmic Pathway have taken 3 billion years to reach Earth.

became hot enough to ...ions, our Sun was born. ...of gas, dust, and icy ...o form the Earth and ...our solar system.

Heavy Bombardment of Planets

Countless comets and asteroids bombarded the inner planets of our solar system during its first few hundred million years.

Big Bang   Present

3 billion years ago
10 billion years since the Big Bang

This unusual radio galaxy has double lobes of gas with plumelike extensions. The overall size spans a distance of more than 360,000 light years. Two jets from the center of the galaxy extend into the lobes.

On the largest scales of the universe, galaxies, clusters of galaxies, and

The universe is all the matter, energy, and space that exists. We can observe only a part of it. The observable universe contains as many as 100 billion galaxies, and because it is 13 billion years old, it extends 13 billion light-years from us in every direction. The entire universe, including the part we cannot see, may be infinite.

Almost all the information we have about the universe comes to us in the form of light, which emanates from stars and galaxies so far away that even with the best telescopes, most appear as mere points. To learn about the universe, astronomers analyze this light in exacting detail.

Light is radiant energy. It travels in waves through space at 300,000 kilometers (186,000 miles) per second. Light comes in many wavelengths. Humans see only visible light, the familiar colors of the rainbow. However, many other "colors" exist beyond the visible spectrum. Radio waves, X rays, and other types of light are the same as visible light but have different wavelengths, changing the "colors" we receive. The wavelength of light determines its energy—the shorter the wavelength, the greater the energy.

IMAGES OF THE NIGHT SKY IN VARIOUS WAVELENGTHS

RADIO WAVES

The longest wavelengths of light are radio waves. The familiar AM and FM radio signals are two specific bands of radio waves, as are television, cellular phone, and satellite signals.

Wavelength range: approximately 1 centimeter and larger

MICROWAVES

The microwave part of the spectrum includes the wavelengths absorbed by simple molecules like water or carbon monoxide. A microwave oven heats food mainly by emitting energy in wavelengths that water absorbs. This energy heats the water in the food.

Wavelength range: 0.1 millimeter to 1 centimeter

INFRARED LIGHT

Infrared light is associated with warmth and thermal radiation. Objects at room temperature, including humans, radiate most strongly in the infrared range. Many "night vision" devices use infrared detectors to track bodies by their thermal radiation.

Wavelength range: approximately 700 nanometers to 0.1 millimeter (1 nanometer = 1 billionth of a meter)

VISIBLE LIGHT

Visible light contains all the colors of the rainbow: red, orange, yellow, green, blue, and violet. The Sun emits mainly visible light. Cooler stars radiate more infrared light; hotter stars radiate more ultraviolet light. If Earth orbited a different star, our eyes would likely have evolved to respond to a slightly different range of wavelengths.

Wavelength range: approximately 400 nanometers to 700 nanometers

ULTRAVIOLET LIGHT

The ultraviolet part of the spectrum has shorter wavelengths and more energy than visible light. By blocking out harmful UV light, sunglasses and sunscreen protect your eyes and skin.

Wavelength range: approximately 10 nanometers to 400 nanometers

X RAYS

X rays are familiar in hospitals. The high-energy X-ray light passes easily through flesh but is absorbed by bones. The film held behind the body part turns black where exposed to X rays and provides a shadow image of the bones.

Wavelength range: approximately 0.1 nanometer to 10 nanometers

GAMMA RAYS

Gamma rays are the highest-energy part of the spectrum. Their energy can do considerable damage to living tissue, so it is fortunate that Earth's atmosphere absorbs gamma rays. Radioactive materials, such as uranium, can emit gamma rays and must be handled carefully.

Wavelength range: smaller than 0.1 nanometer

This ultraviolet image of Saturn shows the planet's bright auroral displays at its north and south poles.

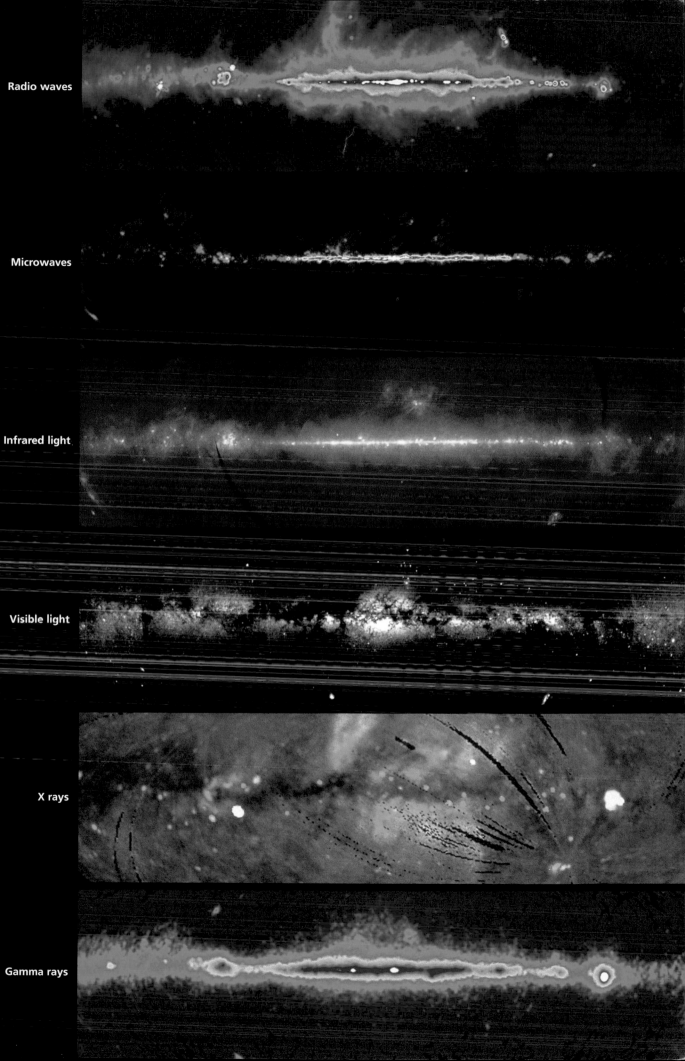

Radio waves

Microwaves

Infrared light

Visible light

X rays

Gamma rays

Galaxies, the basic building blocks of the universe, are titanic swarms of tens of millions to trillions of stars orbiting around their common center of gravity. Galaxies also contain gigantic clouds of interstellar gas and dust strewn among the stars. Even in a tiny patch of sky, many hundreds of galaxies can be seen. Astronomers estimate that there may be 100 billion galaxies in the observable universe.

Galaxies, which are grand yet fragile structures, come in a range of shapes that astronomers group into three basic classes: spiral, elliptical, and irregular.

Below: The Milky Way, home to Earth, is a barred spiral galaxy containing over 100 billion stars. It is approximately 10 billion years old. The spiral disk of stars, gas, and dust is about 100,000 light-years across and only a thousand light-years thick—it's shaped like a pancake. The central bulge of stars is slightly elongated in the shape of a bar. Because we live within the galaxy's disk, we see it as a band of faint light across the sky.

Spiral galaxies have three visible parts:
(1) a thin disk composed of stars, gas,
and dust, (2) a central bulge of older
stars, and (3) a spherical halo of the
oldest stars and massive star clusters.
The signature of these galaxies is an
elegant spiral pattern in the disk.

Messier 100, a grand-design spiral
galaxy, with billions of stars, is about
56 million light-years away.

Elliptical galaxies have smooth, rounded shapes. They contain little gas and dust and no young stars. Unlike spiral galaxies, the shapes of elliptical galaxies are not dominated by motion in a common plane. The orbits of stars in elliptical galaxies are oriented in all directions. Elliptical galaxies, like spirals, are surrounded by halos of globular clusters and dark matter.

Right: M32, a companion to the Andromeda galaxy.

Irregular galaxies have a chaotic appearance and are usually small. Their shapes are probably due to recent disturbances—either bursts of internal star formation or gravitational encounters with external galaxies.

Right: The Large Magellanic Cloud, a companion galaxy orbiting the Milky Way.

Stars are born, live out their lives, and die. An ordinary star is a massive sphere of luminous gas, mainly hydrogen and helium. Its heat and light derive from nuclear fusion in its core. During most of its life, a star is balanced between the inward force of its gravity and the outward pressure of its internal heat.

Stars are born in groups within huge gas clouds called star-forming regions, active places where gas clouds collapse, new stars are born, and aging stars die in spectacular explosions. Between its life and death, the appearance of a star changes dramatically.

The Orion Nebula, a vast interstellar nursery.

In their old age, after depleting the hydrogen in their cores, sufficiently massive stars contract and heat up, triggering the fusion of helium into heavier elements. The increase in energy production puffs out the star's atmosphere, resulting in a highly luminous red giant star. Betelgeuse, a red super giant star in the constellation Orion.

When an intermediate-mass star exhausts its fuel, its atmosphere is blown away and the star's luminous core—a white dwarf—is revealed. It no longer produces energy, and its remaining heat slowly radiates away. The Spirograph Nebula is the expelled atmosphere of a dying star with a white dwarf at center.

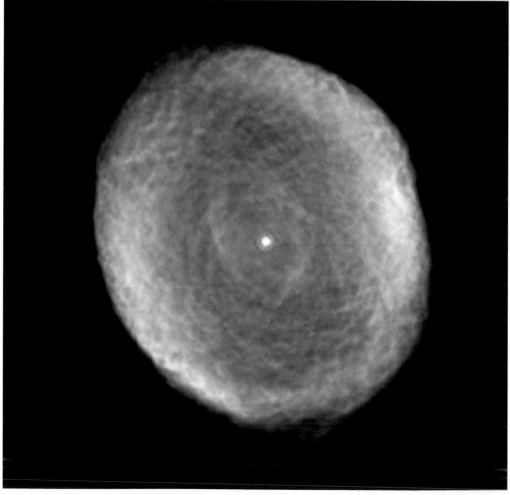

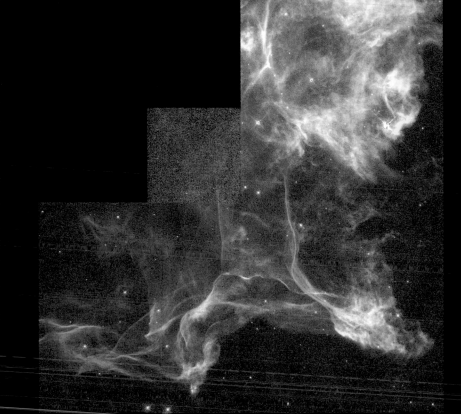

A supernova is the explosion of a star. Runaway nuclear processes at the end of a star's life can cause it to blow itself apart. The core becomes a superdense object—a neutron star or a black hole. The rest of the star is expelled in an outburst brighter than a billion suns. The star's material blasts across interstellar space, forming intricate shells and wisps of glowing gas. Over billions of years, the enriched material ejected by stellar winds and supernova explosions is recycled and becomes part of vast interstellar clouds of gas and dust, from which new generations of stars are formed. Every atom of oxygen in our lungs, carbon in our muscles, calcium in our bones, and iron in our blood was created inside a star before Earth was born.

Left: In the constellation Cygnus, shockwaves from a supernova that exploded 20,000 years ago can still be seen.

Our Sun is an ordinary star, just one among hundreds of billions in the Milky Way galaxy. The closest star to Earth—all others are more than 250,000 times farther away—it is the only one we can observe in detail, and thus it provides a basis for our understanding of all stars.

The Sun is composed almost entirely of hydrogen and helium gas. Its visible surface, called the photosphere, radiates at 5,500 degrees C. Below it lies a turbulent region where the temperature reaches 2 million degrees. This region, which extends a third of the way in, contains two-thirds of the Sun's volume but only 5 percent of its mass. Most of the Sun's mass lies in a stable underlying zone, where temperatures rise to 10 million degrees. In its dense core, temperatures reaching 17 million degrees sustain nuclear fusion, providing the energy source for sunlight. The core contains a third of the Sun's mass packed into 1 percent of its volume.

The internal structure of the Sun.

A prominence of hot plasma in magnetic loops on the surface of the Sun.

The Sun's gravity holds in orbit a family of planets, moons, asteroids, and comets—our solar system. Until the end of the twentieth century, ours was the only planetary system known, but we now know of more planets outside our solar system than within it. Because we cannot observe other planetary systems in detail, there is no universal definition of a planet. In our solar system, planets are the major bodies orbiting the Sun. In general, planets are bodies massive enough for their gravity to make them spherical but small enough to avoid nuclear fusion in their core.

Orbiting our star, the Sun, are the inner rocky and metallic "terrestrial" planets—Mercury, Venus, Earth, Mars—and the outer gas giant planets—Jupiter, Saturn, Uranus, Neptune—along with their many moons. Beyond the outer planets is the Kuiper Belt, a disk of small icy worlds that includes Pluto. Previously considered an oddball planet, Pluto is similar in size, composition, and orbit to the hundreds of recently discovered objects in the Kuiper Belt. The solar system also includes thousands of rocky and metallic asteroids and trillions of icy comets.

TERRESTRIAL PLANETS

MERCURY

Mercury is the smallest terrestrial planet, only 40 percent larger than the Moon. Mercury is close to the Sun, but it rotates slowly and has no atmosphere to hold and transport heat.

Diameter (Earth=1)	0.382
Surface Gravity (Earth=1)	0.38
Surface Composition	Oxygen, Silicon, Iron
Atmosphere	trace
Temperature Range	-180° to 450° C
Period of Rotation	58.6 days
Period of Revolution	88.0 days
Mean Distance from Sun (Earth=1)	0.387

VENUS

Venus is practically Earth's twin in size, but it is radically different in climate. Atmospheric pressure is ninety times greater, and surface temperatures reach 480 degrees C. The surface of Venus, hidden completely beneath a cloud cover, is dominated by ancient volcanic flows.

Diameter (Earth=1)	0.949
Surface Gravity (Earth=1)	0.91
Surface Composition	Oxygen, Silicon, Aluminum, Magnesium, Calcium, Iron
Atmosphere	Carbon Dioxide, Nitrogen
Temperature Range	480° C
Period of Rotation	243 days
Period of Revolution	225 days
Mean Distance from Sun (Earth=1)	0.723

EARTH

Liquid water covers more than two-thirds of Earth's surface. Its rocky crust is broken into about a dozen slowly moving plates. Plate boundaries are marked by seismic faults and great lines of volcanoes. Life on Earth's surface maintains an oxygen-rich atmosphere.

Diameter	12,800 kilometers (7,900 miles)
Surface Gravity	9.8 meters per second per second
Composition	Oxygen, Silicon, Aluminum, Iron
Atmosphere	Nitrogen, Oxygen
Temperature Range	-15° to +40° C
Period of Rotation	23 hours 56 minutes
Period of Revolution	365 days
Mean Distance from Sun	150 million kilometers (93 million miles)

MARS

The red color of Mars comes from iron oxides (rust) in the rocks and dust that cover the planet. Mars has a thin carbon dioxide atmosphere with less than 1 percent of the pressure on Earth. Its polar ice caps of frozen water and carbon dioxide grow and shrink with the seasons.

Diameter (Earth=1)	0.532
Surface Gravity (Earth=1)	0.38
Composition	Oxygen, Silicon, Iron
Atmosphere	Carbon Dioxide
Temperature Range	-120° to +20° C
Period of Rotation	24 hours 37 minutes
Period of Revolution	1.88 years
Mean Distance from Sun (Earth=1)	1.52

Left to right: Mercury, Venus, Earth, Mars.

Jupiter (left) and Saturn.

Neptune (top) and Uranus.

GAS GIANT PLANETS

JUPITER

Jupiter is the largest planet, with a diameter ten times greater than Earth's, but still only one-tenth that of the Sun. Jupiter contains 70 percent of the mass of all the planets combined. Its colors are due to chemical differences between cloud layers at different depths in the atmosphere.

Diameter (Earth=1)	11.2
Surface Gravity (Earth=1)	2.54
Composition	Hydrogen, Helium
Cloud Temperature (at 1 bar)	-110° C
Period of Rotation	9 hours 56 minutes
Period of Revolution	11.9 years
Mean Distance from Sun (Earth=1)	5.20

SATURN

Saturn, the spectacular ringed planet, is almost as large as Jupiter but has only one-third the mass. It has the lowest average density of any planet—only about 70 percent that of water. Saturn has the deepest atmosphere and the highest wind speeds in the solar system.

Diameter (Earth=1)	9.45
Surface Gravity (Earth=1)	1.07
Composition	Hydrogen, Helium
Cloud Temperature (at 1 bar)	-130° C
Period of Rotation	10 hours 39 minutes
Period of Revolution	29.4 years
Mean Distance from Sun (Earth=1)	9.54

URANUS

The rotation axis of Uranus is tipped nearly into the plane of its orbit, so the planet's poles experience decades of constant sunlight alternating with decades of constant darkness. Its pale blue atmosphere consists mainly of water, methane, and ammonia.

Diameter (Earth=1)	4.01
Surface Gravity (Earth=1)	0.90
Atmospheric Composition	Hydrogen, Helium, Methane
Cloud Temperature (at 1 bar)	-195° C
Period of Rotation	17 hours 14 minutes
Period of Revolution	83.7 years
Mean Distance from Sun (Earth=1)	19.2

NEPTUNE

Neptune lives at the edge of night. From its surface the Sun would appear as just another star, albeit the brightest. Neptune's atmosphere displays shades of blue from methane gas, accented with white clouds of methane ice crystals.

Diameter (Earth=1)	3.88
Surface Gravity (Earth=1)	1.14
Atmospheric Composition	Hydrogen, Helium, Methane
Cloud Temperature (at 1 bar)	-200° C (avg)
Period of Rotation	164 years
Period of Revolution	16 hours 7 minutes
Mean Distance from Sun (Earth=1)	30.1

The more than sixty moons in the solar system are a diverse collection of worlds. Two are larger than the planet Mercury. Some of them have ancient cratered surfaces; others are geologically active. Many smaller moons appear to be asteroids or comets captured into orbit by the gravity of their planet.

Jupiter's moon Io, yellow from sulfur and molten silicate rock.

The Moon

The Moon is a dead world. It has no atmosphere or water, so it cannot support life. The mountain-building and erosive forces that have shaped the Earth do not exist on the Moon. Instead, the Moon's surface has been shaped largely by impacts. From Earth we can see only one side of the Moon. Now you can also investigate the hidden side.

Near Side

1. OCEANUS PROCELLARUM
2. MARE IMBRIUM
3. MARE TRANQUILLITATIS
4. MARE SERENITATIS
5. COPERNICUS
6. TYCHO
7. ARCHIMEDES
8. PLATO
9. MARE CRISIUM
10. MARE NUBIUM
11. MARE HUMORUM

Far Side

1. MARE ORIENTALE
2. MARE SMYTHII
3. TSIOLKOVSKY
4. MENDELEEV
5. KOROLEV
6. APOLLO
7. HERTZSPRUNG

View of Moon as photographed from Earth

Shaded relief map from satellite data

Dark areas
The dark areas, or maria (Latin for "seas"), were formed when lava from early volcanoes flooded the enormous depressions created by impacts.

Light areas
The light areas are the highlands, remnants of the original crust that formed when the Moon began to solidify, about 4.5 billion years ago.

Craters
The Moon's surface is pockmarked by countless impact craters, and is covered by a layer of dust and rock fragments ejected from these craters.

Earth's Moon is a dead world. It has no atmosphere or water. The mountain-building and erosive forces that shaped Earth do not exis on the Moon. Instead, the Moon's surface has been shaped largely by impacts. From Earth we can see only one side of the Moon

Opposite: This aquatic environment is a microcosm of a living world. The sealed sphere contains a complete ecosystem of plants and animals that recycle nutrients and obtain energy from sunlight. A planet that harbors life, like Earth, must maintain a similar balance.

Above: The Willamette meteorite was found 3 km (1.8 miles) northwest of the village of Willamette (now West Linn), Oregon, in 1902. A metallic iron meteorite weighing 15.5 tons, it is the largest found in the United States and one of the largest in the world. Its cone shape is a result of ablation—the removal of material by heating as it entered the atmosphere. The large holes on its flat surface formed over the thousands of years that the meteorite has been on Earth. Rainwater interacting with deposits of iron sulfide in the meteorite produced sulfuric acid, which created the cavities.

The range of size scales between the largest and smallest objects in the cosmos is truly astronomical. The observable universe is a billion billion times bigger than Earth; humans are a million billion times larger than a proton in the nucleus of an atom.

Left: The Hayden sphere, 87 feet in diameter, is the reference point for the Size Scales of the Universe. When the sphere represents the Sun, for example, Earth would be only 10 inches across, meaning that more than a million Earths could fit inside the Sun.

Above: If the Hayden sphere is the size of a raindrop (2 millimeters), then the relative size of a red blood cell is 8 micrometers. More than 10 million red blood cells are in a drop of blood the size of a raindrop.

WHY ARE THERE OCEAN BASINS, CONTINENTS, AND MOUNTAINS?

Over millions of years, ocean basins open and close, continents move and change shape, and mountains are pushed up and eroded away. Such dynamic processes continually reshape the surface of the Earth. The movement of rigid plates on the Earth's surface, known as plate tectonics, is the cause of these changes. Volcanic eruptions and earthquakes are dramatic hints of the great movements that take place over the vastness of geological time.

THE DAVID S. AND RUTH L. GOTTESMAN
HALL OF PLANET EARTH

Earth is constantly changing, reworking itself into new and complex forms. Clouds materialize and melt away; wind and water and ice carve the land into new shapes; and the slow churnings of Earth's interior keep the continents in motion and create new oceans and mountains. The interactions of the atmosphere, ocean, and solid Earth and its flora and fauna make the planet dynamic and have shaped and continue to shape its evolution.

In order to understand these geological processes, we need to envision time periods far greater than the hourly, daily, and yearly progression that dominates our lives. Major changes in the continents, oceans, atmosphere, and biosphere occur over millions of years. But Earth's rocks provide clues that allow scientists to piece together the planet's 4.5-billion-year history.

The Hall of Planet Earth is dedicated to displaying and exploring the way our dynamic Earth works.

The half-globe suspended above the hall—Earth as seen from space—is composed from the latest satellite imagery. Eight feet in diameter, it shows varying views of the planet. Clouds peel back to reveal oceans and continents. Then the land is stripped of vegetation, and water drains from the oceans, exposing deeper and deeper levels of ocean floor. Finally, Earth appears without atmosphere and water.

Over millions of years, ocean basins open and close, continents move and change shape, and mountains are pushed up and eroded away. The movement of rigid plates on Earth's surface—known as plate tectonics, one of the grand unifying theories of geology—is the cause of these continual changes and connects seemingly unrelated features and events of the planet—continents and oceans, mountains, volcanoes, and earthquakes—to a single global process. These rigid plates are constantly being formed, altered, and consumed. Moving on the mantle below them, they carry the continents along with them.

Earth's crust is just a thin veneer lying on top of the mantle, the middle layer of Earth's three components. The lighter rocks of the continental crust float higher on the mantle than the denser rocks of the oceanic crust. The relief of Earth's surface—the difference between the deep ocean basins and the high continents—is caused by this difference in density.

For emphasis and clarity, the surface detail on the Wallace Gilroy bronze model Earth is vertically exaggerated 22.5 times.

Right: The mountain ranges that span the globe mark areas where Earth's tectonic plates converge. Pieces of the crust are piled on top of one another, creating complex patterns of folds and faults. Some geologists build models with sand, since the way sand moves mimics the way rocks are slowly changed by high temperature and high pressure over geological time.

Left, top: This sample of garnet-bearing amphibolite from the Adirondack Mountains in New York is a spectacular illustration of how a rock can undergo a complete change in texture through metamorphism—the transformation of one rock to another by heat, pressure, and contact with reactive fluids—and through deformation—the change in the form of a rock under high temperature and pressure. Originally, this rock was a gabbro consisting of tiny grains of the minerals olivine, pyroxene, and plagioclase. During deformation, water seeped into the gabbro and reacted with the original minerals to form hornblende (black) and garnet (red). The water allowed the minerals, especially the garnet, to grow to enormous sizes.

Left, center: From Zermatt, Switzerland, this eclogite represents a sliver of oceanic crust that became trapped and caught up in the continental collision that formed the Alps. The red garnet and spinach-green clinopyroxene indicate that the rock recrystallized at depths of 50 to 60 kilometers (31 to 37 miles) and at temperatures of about 600 degrees C. The white mineral, epidote, formed later, during uplift and cooling.

Left, bottom: When rocks are compressed, they deform by bending and folding. When a conglomerate—a rock composed of rounded pebbles surrounded by sand or silt—is subjected to high temperature and pressure, the pebbles within it are deformed in different ways. They can be flattened, rotated, and even squeezed into long, thin lines. This rock from the Hemlo greenstone belt in Ontario was originally deposited in a stream as layers of rounded pebbles. As these layers were eroded and redeposited, they became mixed with sands and silts. Much later, the rock was transformed by high temperatures and pressures deep in the crust, and the pebbles were stretched, flattened, and folded.

Until the end of the eighteenth century, scientists believed that Earth was no more than a few thousand years old. They claimed that catastrophes were the prime sculptors of the planet. Fossils were thought to be the remains of animals that had perished in the biblical Flood, when sedimentary rocks had also formed.

The Scottish naturalist James Hutton (1726–1797) began to formulate geological principles based on observations of rocks. A key site was Siccar Point, a sea cliff east of Edinburgh where horizontal layers of red sandstone rest on near-vertical folded layers of gray slate and sandstone. Hutton concluded that the gray rocks had been deposited horizontally, then uplifted, folded, tilted, eroded, and again covered by the ocean, from which the overlying red sandstone accumulated. He recognized that these processes must have taken a very long time. The boundary between the two rock sequences is called an unconformity. The unconformity represents the period of time when the underlying gray rocks were eroded, before the red rocks were deposited.

Above: Siccar Point.

Opposite: Museum workers make a latex mold of the Hutton Unconformity to reproduce in the Hall of Planet Earth.

Right: The sediments that formed these Old Red Sandstone specimens from Siccar Point (right) and Monticello, New York (left), were laid down in the same basin during the Devonian period, about 400 million years ago, when they shared one vast supercontinent. When the supercontinent broke apart into the continents we see today, the sandstones were separated. Small differences in grain size and in the composition of the material that cements the grains together account for their different color and texture.

There are three types of rock: igneous, sedimentary, and metamorphic.

Igneous rock forms when molten rock (magma or lava) cools and solidifies. It is identified by the size and shape of its mineral grains. Igneous rocks' texture is controlled by the rate of cooling: magma, which cools slowly deep in the Earth, forms rock with large crystals, and lava, which cools quickly on the surface, forms fine-grained rock.

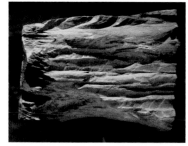

Igneous

Sedimentary rock is created in layers when particles of rocks and minerals settle out of water or air or when minerals precipitate directly out of water. As these sediments pile up, water is driven out by the weight of the overlying pile, and minerals precipitate around the sediment particles, cementing them into rock in a process called lithification.

Sedimentary

When sedimentary, igneous, or preexisting metamorphic rocks are changed by heat, pressure, or chemically reactive waters into new minerals with different sizes, shapes, and orientations, they are called *metamorphic* rocks. Most metamorphic rocks are made of minerals containing silicon and oxygen, the most abundant elements in Earth's crust. They are identified by differences in their mineral content and texture.

Metamorphic

This granite pegmatite (top right) from the Ehrenfriedersdorf Mine in Germany formed deep in Earth's crust, near the top of a crystallizing magma chamber. It is composed of the minerals quartz (gray), orthoclase (pink), albite (white), and mica (dark and platy). The larger crystals grew more slowly than the smaller crystals. This diabase (bottom right) from the Palisades Sill, Haverstraw, New York, solidified from a basaltic magma within a few hundred meters of the surface, probably beneath a volcano. It cooled rapidly, giving it a fine-grained, peppery appearance. The black mineral is pyroxene, and the white one is plagioclase. This diorite (left) from Rapidan, Virginia, crystallized in a mountain belt. It contains the minerals plagioclase (white) and hornblende (black).

Limestone is one of the most widespread sedimentary rocks. Many organisms, from corals to microscopic foraminifera, grow shells composed of carbonates. Most limestone, like this from near Albany, New York (left), forms when these organisms die and their shells accumulate in shallow seas. Shale (top right) is made up of clay and silt, which are deposited in slow-moving rivers, at the far ends of river deltas, and in other quiet environments where water cannot keep the particles suspended. This specimen is from Sedona, Arizona. Sandstone is made of sand, which accumulates in rivers and along ocean shores where water moves quickly enough to remove finer particles. Sandstone can also be deposited by winds. This specimen (bottom right) is from Utah.

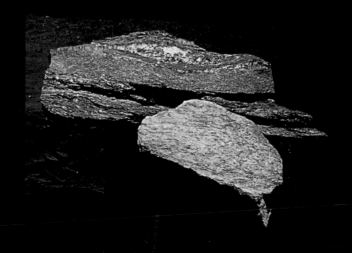

Originally a shale, this Manhattan schist from Central Park in New York City (top) now consists of dark, biotite-rich layers and finer-grained, light-colored layers made of gray quartz and white orthoclase. The distinct layering developed perpendicular to the direction of compression. Manhattan schist underlies most of Manhattan Island. This gneiss (bottom) from the Adirondack Mountains, near Elizabethtown, New York, was once a granite. As it was deformed by heat and pressure, the original even-grained rock developed dark layers, primarily biotite, and light layers, mostly gray quartz, pink orthoclase, and black hornblende.

This banded iron rock records a time from Earth's distant past when early life profoundly influenced the planet's evolution. The abundance of oxygen that is now in Earth's atmosphere was not there at the beginning. Early life began to generate oxygen, by converting the Sun's energy into food. This caused the iron that was dissolved in the oceans to precipitate out as iron oxide minerals. This rock, with its layers of red jasper and iron magnetite, was formed billions of years ago as part of this process.

By dating zircons embedded within the Acasta gneiss, geologists have determined its age to be 3.96 billion years—which makes this the oldest known rock. Its composition establishes it as part of a continent, indicating that continents existed that long ago.

The cross-bedded Coconino Sandstone (upper sample) from Sedona, Arizona, formed as a windblown sand in a desert environment. Prominent layers are visible, and within them are thinner inclined layers known as cross-beds. They indicate the direction of the wind that transported the sand. The upper surface of the sandstone from Elliot Lake, Ontario (lower sample), shows a wavy pattern known as ripple marks. Such marks are created by the movement of water or wind over sediment and indicate the direction of the current and orientation of the surface when the sediments were deposited.

This gray granite rock from Elberton, Georgia, solidified between 350 and 320 million years ago. More than 120 million years later, molten basaltic rock forced its way up through a fracture in the granite, forming this gray-black dike. The margins of the dike, which cooled quickly against the cold granite, are finer grained than the interior, which cooled more slowly, allowing crystals to grow larger.

Geologists peer into Earth using a seismometer, an instrument that senses ground movements from earthquakes. When seismic waves cause ground motion, its rotating drum, or seismograph, graphically records the movement over time.

Opposite: Most earthquakes occur at tectonic plate boundaries; rocks break, forming fractures called faults. A fault is a rock fracture along which movement occurs. Normal faults develop where Earth's crust stretches apart. This near-vertical fault in granitic gneiss from Essex County, New York, offsets a black basaltic dike.

There are two types of volcanoes, explosive and effusive.

Most explosive eruptions occur in volcanoes above subduction zones, where one tectonic plate dives beneath the other. Magma—molten rock—forms 80 to 120 kilometers (50 to 75 miles) below the surface, when the rocks of the mantle melt just above the subducting plate. The magma rises through the mantle and erupts to form explosive volcanoes like the 127 active ones in Indonesia. Volcanic gases are composed mainly of water, carbon dioxide, and sulfur dioxide.

Kawah Ijen volcano, Java.

Glacial ice records the climate of past ages. Each year, as old snow is buried under new, it compacts and recrystallizes into layers of ice. The composition of the ice itself and of the air bubbles and dust trapped in it capture changes in temperature, humidity, atmospheric circulation, volcanic activity, extent of sea ice, and even atmospheric pollution by human activity. At the top of the Greenland Ice Cap in Summit, Greenland, glaciologists have drilled through 3,022 meters (9,915 feet) of accumulated ice, all the way to bedrock. The 1-meter (39.37-inch) ice cores they extracted contain a detailed climate record, extending back 115,000 years.

Simulated ice core samples. Left: depth, 2,708 meters (8,884 feet); date, 88,670 B.C.; middle: depth, 1,689 meters (5,541 feet); date, 10,075 B.C.; right: depth, 1,325.5 meters (4,349 feet); date, 5671 B.C.

Right: This piece of coral (*Porites lobata*) from Champion Island, in the equatorial Pacific Ocean, like all corals, is a skeleton built of calcium carbonate extracted from seawater by coral polyps; it grew from approximately 1740 to 1900. Because coral grows near the ocean surface, it is a useful indicator of local past climate. The thickness of the layers, which varies in density from summer to winter, reflects changes in temperature, salinity, and cloud cover.

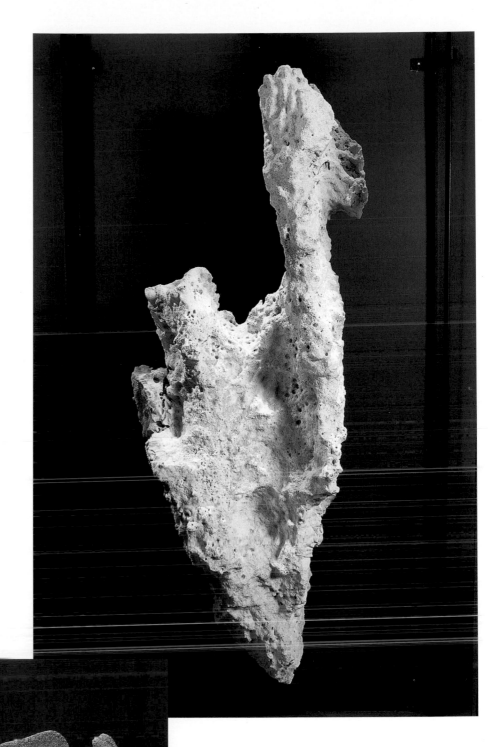

Below: This ventifact, or wind-carved rock, from Bull Pass in the Taylor Valley, Dry Valleys of Antarctica, demonstrates how rocks can be shaped into sculpturelike forms by erosion and weathering. The primary erosion comes from wind abrasion. Chemical weathering, expansion as a result of heating, and contraction as a result of cooling also contribute.

An unusual environment for life exists deep in the oceans. Where hot springs emerge on the ocean floor, a microbial community flourishes, living off the hydrogen sulfide and other compounds carried by the venting fluids. Some of these microbes represent the most ancient life known, and they form the base of a food chain for a group of organisms that never see the light of the Sun. Life may have begun around these deep hydrothermal vents. If similar environments exist elsewhere in the solar system, they too may support life.

Such an environment lies at the Juan de Fuca Ridge, 2 kilometers (1.6 miles) under the northeastern Pacific Ocean, and elsewhere along the global system of midocean ridges. Heat from intrusions of magma causes seawater to circulate through the cracks in the rocks around these vents and dissolves metals like iron and copper out of the rocks. Where the hot water returns to the ocean, tiny particles of metal sulfide and oxide minerals resembling black smoke billow from the vents. As the hot water flows from these "black smokers," these particles begin to build chimney-like structures with organized honeycomb-like channelways. As the chimneys grow, their interiors become insulated from the cold seawater, allowing zones of high-temperature minerals to form along the channelways.

In 1997 and 1998, the American Museum of Natural History and the University of Washington sent an expedition to the Juan de Fuca Ridge to bring back several active chimneys, to study how they grow, and to determine the nature and distribution of life inside them. These three chimneys from the Mothra hydrothermal field are named for characters in Irish mythology: the warrior Finn MacCool; Gwenen, Guinevere's lusty sister; and Roane (Gaelic for "seal"), one of the fairy people.

Opposite: Motha Hydrothermal Field, Endeavor Segment, Juan de Fuca Strait. Chimney Finn is third from the right, and Chimney Roane is second.

Left: These are two halves of the top of the Chimney Gwenen (left), a five-story-high spire that was venting water at 179 degrees C. It bristled with a profusion of tubeworms and other animals, vestiges of which can still be seen on its surface. As the chimney grew, built up irregularly by minerals precipitating from the water, animals were trapped and buried in the walls.

The Chimney Roane (right) contains a complex series of interconnected channels and isolated voids. Some of the channels are lined with the yellow metallic copper-iron sulfide chalcopyrite, which forms only at temperatures approaching 300 degrees C. The temperature of the water seeping from the top of Roane was 196 C.

The water flowing through the Chimney Finn (behind the Chimney Gwenen) was measured at 302 degrees C. After the chimney was collected, the stump grew nearly 3 meters (10 feet) in three weeks, owing to the large amount of hot metal-laden water flowing through it.

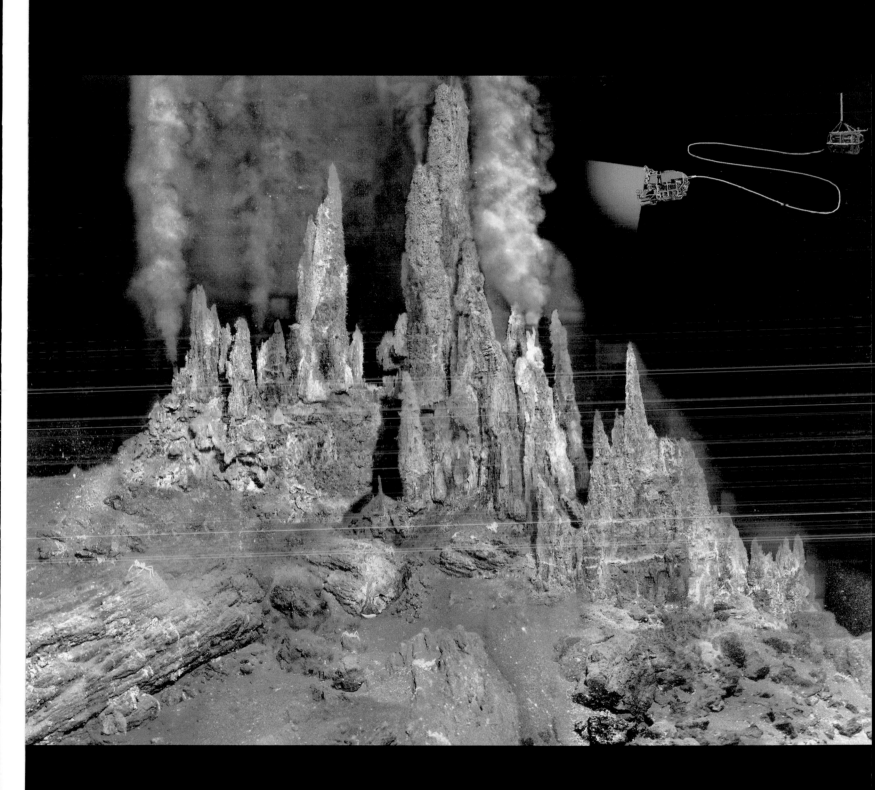

BUILDING THE ROSE CENTER

AMY WEISSER

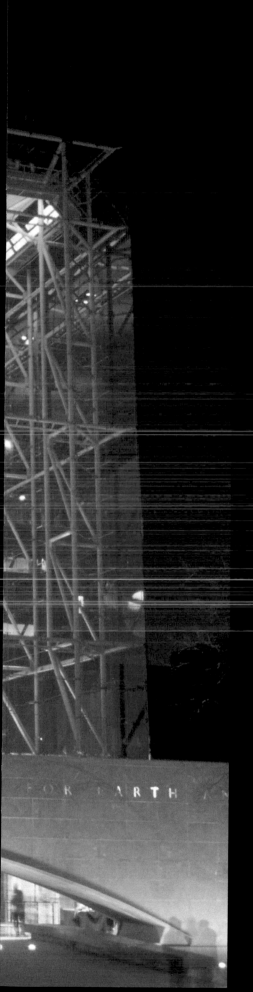

Standing five stone-clad stories tall and ornamented with turrets, towers, and columns, the American Museum of Natural History presents to the world an august public institution with roots in the nineteenth century. But walk past the carriage entrance that cleaves the Museum's 1891 facade on 77th Street and the broad flight of steps of the 1936 facade along Central Park West, and turn the corner. There a twenty-first-century museum awaits you. The heavy walls of the adjacent exhibition wings give way to sheer planes of glass enclosing a gleaming aluminum-covered sphere—this is the American Museum of Natural History's new Rose Center for Earth and Space.

The transparent envelope of the Frederick Phineas and Sandra Priest Rose Center for Earth and Space—124 feet wide and 95 feet high—reveals an astronomical wonderland inside.

The 87-foot-diameter sphere that serves as its centerpiece is surrounded by several colorful smaller spheres, including one depicting Saturn with rings measuring 17 feet across, and a 9-foot-wide Jupiter. Even before reaching the entrance, you realize—first—that the seemingly floating, luminous sphere could represent the Sun, and—second—that where you really want to be is inside the building.

By making clear that what lies within has serious scientific content, the Rose Center for Earth and Space fulfills the Museum's first ambition for its architecture: This is architecture in the service of science. "The real goal [of the Center]," says the president of the Museum, Ellen V. Futter, "is to make this exceedingly complex, abstract area of science accessible and comprehensible to the public, to bring the frontiers of outer space and discovery to the people and, in so doing, to illuminate the magnitude, the majesty, and the mystery of the cosmos."

The Rose Center, a 207,000-square-foot research and exhibition facility, houses a complex array of facilities and programs:

- the new Hayden Planetarium (the giant sphere) consisting of the Space Theater and a performance space featuring a simulation of the Big Bang
- two new permanent exhibition halls, the Cullman Hall of the Universe and the Gottesman Hall of Planet Earth
- a one-acre public terrace
- six stories of support facilities, with classrooms for the Museum's educational programs and offices for the Department of Astronomy and Astrophysics.

In addition to these exhibition and educational facilities, the Museum has added a parking garage, dining facilities, and a new entrance. Polshek Partnership Architects designed the Rose Center, with Ralph Appelbaum Associates as the exhibition designer.

Entering the Rose Center through Theodore Roosevelt Park, which borders the Museum on its north face, you walk through a stone archway whose density provides a visual contrast to the glass above, acting as a kind of visual anchor for the ethereal superstructure while its gentle curvature recalls the Museum's southern entrance along 77th Street. Once inside the public foyer, you approach an overlook. Here you glimpse the underbelly of the sphere but see none of its supports. The giant ball seems to hang weightless in the larger volume of space. Looking out, up, and down, you see the walkway that surrounds the sphere, the orbs suspended from the ceiling, and the clusters of displays in the Hall of the Universe below. The globe is the icon for the entire center and a core element of its exhibits: It teaches that the

universe is filled with spherical objects so formed by the forces of gravity and energy and serves as a reference point in an exploration of scale relationships with objects from atoms to planets to stars to galactic holes.

Both iconic form and primary exhibit, the giant sphere also houses the two levels of the Hayden Planetarium. In the upper hemisphere, the Space Theater's 429 high-back seats slope gradually down from the outer wall, and the space rises 38 feet to a hemispherical scrim wall just inside the exterior aluminum cladding. The lower portion of the sphere inverts that architectural experience. Visitors stand over an aperture in the floor while projections against a curved surface visually energize the depths of the sphere. The Big Bang simulation featured there leads directly and thematically to the Cosmic Pathway, a spiraling 13-billion-year timeline some 360 feet in length that coils around the sphere one and a half times on its way to the first floor.

The landing platform, devoted to Earth's moon, is a directional pivot, a bridge to the Rose Center's Hall of Planet Earth, which is on the first floor of the Museum's 1933 Whitney Wing along Central Park West. And it also leads to a grand curvilinear stair that connects to the Hall of the Universe on the lower level.

The Hall of Planet Earth, at more than 8,500 square feet, is a high-ceilinged rectangular space typical of the Museum's traditional exhibition halls in size and shape. Entering the Hall of Planet Earth at the center of one of its long walls, you come first to the hemispherical Dynamic Earth Globe, suspended from the ceiling,

with video imagery of an astronaut's view of this planet. From the Hayden Planetarium sphere to the moon model to the Earth Globe, a celestial continuity is clearly established. In a circumferential arrangement of rock samples, the remainder of the Hall explores the fundamental questions of Earth science. In contrast to the rest of the Rose Center, which is light filled and outward looking, the fully enclosed Hall of Planet Earth, with its dark, rich colors, appears to be situated in the Earth itself.

The Hall of the Universe—reached from the main entrance as well as from the moon stair—is a 7,000-square-foot open space

comparable to that of the glass above. Brick and copper, also primary materials on the exterior, were chosen to recall the existing materials on the Museum's north side.

Among the striking features on the interior are the stainless steel fittings along the glass curtain walls. Delicate, curvaceous, and remarkably spiderlike, these fittings join the multiple panes of glass to the overall structural system supporting the walls. They were cast to exacting, almost sculptural, specifications, and were treated with a "shot-peened" finish for a warm luster. Facing the glass walls, the metal covering the sphere offers a different kind of engineered exactitude. The sphere is clad in 2,474 panels of stretch-formed aluminum placed at 15-degree longitude and latitude divisions. To provide acoustic baffling, the sphere is perforated with 5.6 million tiny holes, each row of which was separately calibrated to accommodate the sphere's curvature. Finally, the sparkling black composite-stone floor in the entrance and other areas is embedded with Czechoslovakian glass shards that twinkle in a rainbow of colors.

The systems that lift and support these materials are captivating in and of themselves. Many of the engineering techniques that make the dramatic architecture possible are in use for the first time in the United States. Moreover, because of the exacting nature of the design, much of the construction was performed at a level far in excess of industry standards. The ironwork that supports the sphere, for example, required precision to within one-sixteenth of an inch of the architect's specifications. The architect and contractor relied on a host of consultants, including Weidlinger Associates, New York City structural engineers; TriPyramid Structures, tension structure consultants and fabricators from Concord, Massachusetts; and Chicago Ornamental Iron, steel fabricators from Chicago. Computer-aided design (CAD) was essential throughout the process, although some good old-fashioned three-dimensional model building played a role, too.

The Rose Center's glass wall is billed as the largest suspended glass curtain wall in the United States. It consists of two glazed areas totaling 22,800 square feet, 8-foot-wide areas along the adjoining two sides of the cube, and a perimeter skylight at the top of the box. This skylight highlights the glass wall's independence from the roof structure. To ensure maximum transparency, the architects developed structural support systems for the glass walls that are as unobtrusive as possible. The resulting

elements—codependent sets of wall trusses and curtain wall tension trusses—
obscure less than 10 percent of the glass wall area. The wall trusses are composed of
a series of vertically arranged tubular steel pipes no more than 8 inches in diameter.
They are connected to pedestals by pins at the floor level to allow for slight movement, while they are rigidly connected at the roof to a series of lateral trusses. Rods
and cables stiffen the system, with additional support from horizontal steel pipes. The
horizontal curtain wall tension trusses are made of high-strength stainless steel,
which is more resistant to wind loads than conventional steel or aluminum. Again,
perfection counts: Over the entire 95-foot height of the glass box, the engineering
specifications demanded that the structure conform to ideal measurements with no
more than three-eighths of an inch to spare.

Whereas most curtain walls use aluminum frames to support individual panes of
glass, at the Rose Center the illusion of a seamless field of glass is maintained by the
use of flush, countersunk bolts to join glass to "spider" fitting to truss system. The
only separation between the panels are half-inch joints of translucent silicone sealant.
The wall system uses 1,400 cast stainless steel "spider" fittings, 4,100 bolts, 2.5 miles
of rod rigging, and 40,000 square feet of glass. Built into the structural system is a
roof hoist for an electric bosun's chair and a 30-foot boom so that the glass wall and
the metal sphere can be easily cleaned.

While the structure of the curtain wall is rational and reductivist, that for the sphere is complex and dramatic. With the amenities and equipment it contains, the sphere weighs 4 million pounds and hovers 23 feet above the Hall of the Universe. Its weight is supported on three V-shaped hollow-steel inclined columns, which resemble the base of a giant tripod. These columns join the sphere at a precise 68.9153-degree angle and taper from 2 feet 7 inches at the ends to 3 feet 9 inches at their middle. They extend 57 feet and attach to the sphere just below its midpoint, supporting a 12-foot-deep ring truss at its equator. This massive steel structural girdle, in turn, provides the spring point for a series of radiating ribs that form the back-bone of the sphere's exterior cladding. The detailing of the structure of the sphere was so challenging that Chicago Ornamental Iron found it beyond the capacity of its CAD system and, instead, preassembled the pieces in its shop.

The tripod columns, supplemented by five smaller legs, also hold up the ramp that corkscrews out of the sphere, transporting visitors from the second floor to the first. The ramp's 8-foot-wide footpath cantilevers from a torsion tube of 30 inches in diameter that forms the path's outer edge. This daring structural system allows the pathway to span up to 90 feet between supports. Even more striking, the ramp curves in two planes, circling 540 degrees around the sphere while dropping 16 feet in altitude. The architect located an induction-bending machine capable of curving the torsion tube to specified radii of up to 51 feet at BendTec, Inc., a pipeline construction company in Duluth, Minnesota.

While the systems supporting the curtain wall, the sphere, and the ramp are captivating in the manner in which they push the envelope of what is possible, the overall effect is to diminish the importance of structure so that its platonic forms and pristine materials come to the fore. Once you are inside the Rose Center, the sphere looms so large in your field of vision that its parameters seem barely perceptible. Moreover, although the three pairs of columns supporting the sphere are visible, the space between them diminishes their Atlas-like role, and the mammoth ball of

metal suggests a floating bubble. The clarity of the architectural armature of the Rose Center brings to the fore such tangible qualities as the play of natural light throughout the day, the delicate color variations of glass, metal, and wall, the simple geometries of space, and the subtle changes in the visitor's experience of this space from level to level.

While the exhibition spaces of the Hayden Planetarium, the Hall of the Universe, and the Hall of Planet Earth are the major foci of the Rose Center, the construction project also facilitated a number of visitor amenities—including the parking garage, dining facilities, retail outposts, new entrance along the Museum's Columbus Avenue facade, and landscaped one-acre terrace—as well as office space and classroom space. Together, all these components finish the Museum's north side, which had been left incomplete when work on the Museum's 19th-century master plan ceased in the 1930s.

The parking garage, which accommodates 370 cars, consists of three stories—two below ground and one at grade level—with the terrace above. The dining facilities are outfitted in the Museum's former power plant, which anchors the northwest corner of the Museum's plot. Retail shops are integrated with the interior architecture on the lower and first floors. The Columbus Avenue entrance is the first pedestrian access from the west for the Museum's Upper West Side neighbors. Forty-three feet high and sheathed in glass, it recalls the monumental glass cube along 81st Street. Theatrically displayed inside is a modern update of a classic armillary sphere, an astronomical instrument composed of concentric rings that illustrate the positions of celestial grids relative to Earth. This instrument depicts New York City's precise celestial location on January 1, 2000.

Directly to the west of the Rose Center cube and entered from 81st Street, the Arthur Ross Terrace, set on a platform 18 feet above the adjoining Theodore Roosevelt Park, provides more than an acre of public space. Planned by landscape designer Kathryn Gustafson in collaboration with the architects and the landscape architecture firm of Anderson and Ray, the terrace uses stone, water, light, and plantings to express a metaphoric astronomical landscape. Its central element, a broad swath of sloping ground, is animated by wedge-shaped areas defined in stone, which increase in size toward Columbus Avenue. These planes represent the shadow cast by the Hayden Planetarium sphere and are reminiscent of the multiple conical shadows cast by planets and moons. The center of the plaza is rendered in a blue-green stone punctuated by jets of water and fiber-optic lights set in a pattern based on the Orion star cluster.

Paths etched into the stone, evocative of meteor trails, draw rills of water to a reflection pool at the base of the glass cube. Periodically, this central field can be flooded in a thin wash of flowing water.

A double row of pagoda, or scholar, trees (*Sophora japonica*) lines a promenade and provides filtered light along the northern edge of the terrace. Underneath the trees, a staggered line of large wooden benches provides seating for school groups and other visitors. Cascading down the brick wall of the garage, beach plum (*Prunus maritima*)

provides bursts of flowers each spring, making a mist of white petals. Opposite, an irregularly curving wall of benches evokes a geologic fracture. At the entry stairs, cultivars of mountain laurel (*Kalmia latifolia*), including Star Cluster, Comet, Shooting Star, and Galaxy, fill planters with star-shaped blooms. Native plants including inkberry (*Ilex glabra*) and franklinia (*Franklinia alatamaha*) refer to the local landscape.

A small upper plaza is shaded by ginkgo trees (*Ginkgo biloba*), the only surviving species of a family that originated during the Carboniferous period (340 to 290 million years ago). In the autumn, when the leaves of these living fossils turn bright yellow, a solar haze is cast on the tables and chairs below.

With a focus on providing a conceptual and aesthetic link between the Rose Center and the surrounding Museum buildings and park, the terrace's simple plan, narrow palette, and precisely rendered plantings balance modernity and tradition. As such, they provide a fitting border between the Museum and its neighborhood and between outdoor space and the natural world.

The architect for the Rose Center and the Museum's entire northside renovation was Polshek Partnership Architects, a New York City firm best known for institutional projects. Design principals for the Rose Center project were James Stewart Polshek, the founder, and Todd H. Schliemann, a partner at the firm. More than forty architects and designers served on the project team, with the assistance of consultants from over twenty-five disciplines.

Ralph Appelbaum Associates worked closely with Museum staff to create the Rose Center's exhibits. The New York City–based firm, founded in 1978, came to national attention with its design for the United States Holocaust Memorial Museum in Washington, D.C. (1993), and has won many awards for its work. Principal Ralph Appelbaum led the team for the Rose Center project.

NEW TECHNOLOGY FOR A NEW MUSEUM

ROBERT J. NELSON

Enclosed in a 95-foot-high cube of gleaming glass, the Rose Center is the newest and most technologically complex of the American Museum of Natural History's forty-two exhibition halls. Dedicated to research and education on astrophysics and Earth science, the center mirrors technologically the rapid pace of scientific research in these fields; every exhibit has been designed to be easily updated when new information becomes available, incorporating multimedia displays that can be reprogrammed, in some cases by downloading the latest data directly from satellite. Thus the Rose Center puts cutting-edge technology into the service of science and education.

THE DYNAMIC EARTH GLOBE

Located in the Hall of Planet Earth, the Dynamic Earth installation presents the sun-lit side of the rotating Earth on an 8-foot-wide overhead hemispherical screen—a stunning view of Earth as it appears from outer space. Signals from a computer in the Museum's second-floor control room cue the succession of scenes. Inside the screen is a computer graphics projector fitted with a custom 180-degree lens that projects on the screen an image that is 1,024 pixels across, with each pixel represent-ing an area of approximately 20 square kilometers (7.7 square miles). During a twelve-minute presentation, the planet revolves twice, presenting different parts of Earth as clouds evaporate, vegetation disappears, and seas recede, exposing the dry ocean basins. The desiccated Earth is then restored as the waters, vegetation, and atmosphere return in reverse order.

The clouds on the screen are based on actual clouds imaged by satellite on September 13, 1996, and much of what visitors to the Museum see is how Earth looked on that day. The cloudless images of Earth are also based on real imagery, derived from a considerable volume of satellite data pieced together to form a com-plete mosaic. For landmasses, data were drawn from a polar-orbiting weather satel-lite operated by the National Oceanic and Atmospheric Administration's (NOAA) Television Infrared Observation Satellite (TIROS). A polar-orbiting weather satel-lite from the U.S. Air Force's Defense Meteorological Satellite Program gave infor-mation on ocean surfaces, sea ice, snow, and cloud cover.

The modelers turned to the United Nations Food and Agriculture Organization's Soil Map of the World to help render land colors. To render ocean color, they relied on the National Aeronautics and Space Administration's Nimbus-7 Coastal Zone Color Scanner program for satellite measurements of phytoplankton, the tiny organ-ism that imparts a greenish hue to the water.

Measurements of the seafloor topography were derived by two scientists, Walter H. F. Smith of NOAA's National Geophysical Data Center in Boulder, Colorado, and David T. Sandwell of the Scripps Institution of Oceanography at the University of California, San Diego, who relied on a combination of shipboard depth soundings, satellite gravity data, and satellite-measured height above the ocean surface.

The computer animation and the Earth layers it incorporates were created by ARC

Science Simulations, a company in Loveland, Colorado, that specializes in computer modeling and display of the visible planet. To produce the Dynamic Earth animation, the modelers first created a computer image of Earth stripped of vegetation, snow, ice, and water. The most difficult work at this stage was creating plausible ocean basin coloration with additional detail that would be brought out by sunlight shining on the seafloor topography. Another challenge was to depict the land stripped of vegetation in plausible color and with detail that would reflect true variations in soil and minerals; it also had to appear reasonable in juxtaposition with deserts and other arid lands. Clouds in the Earth animation don't just fade away; they evaporate, with massive storm systems shrinking away last. Vegetation fades out quickly, though, as might happen on a planet where the atmosphere is drying out.

After this, the oceans recede according to depth. The relatively shallow continental shelves are exposed quickly, and then—since most of the ocean basins are about two miles deep—the receding water appears to slow a good deal as islands, seamounts, and volcanic ridges rise into view. Water remains in the deep ocean trenches marking the subduction boundaries of the tectonic plates. Last to dry up, the Mariana Trench east of the Philippines, which, at 10,920 meters (35,827 feet) is deeper than Mount Everest is high. The illumination of the sun is accurately reproduced, so topographic features both on the land and in the ocean basins are enhanced by deep shadows when they are near the edge of sunlight.

SOUNDSCAPES: THE INVISIBLE ATMOSPHERE

The Hall of Planet Earth was acoustically designed to reinforce the experience of geologic process. Sounds sweep around the Hall, illustrating the scenes displayed on the Views of the Planet screen. In addition, in each exhibit area very local sounds—rock slides, volcanic eruptions, and other Earth events—are synchronized with the displays so the viewer hears as well as sees the display. The antiphonal pings of a geologist's pick are the only projected sounds in the hall made by a living creature.

Three soundscape scripts were installed in computerized 8-track and sixteen synchronized playbacks. Most of those channels are associated with the Views of the Planet video screen at the north end of the Hall. Large subwoofers under the floor and smaller ones mounted under benches on which visitors may relax generate deep bass tones below the range of human hearing, giving visitors the visceral sense of vibrations produced during earthquakes, landslides, and volcanic eruptions.

The plate tectonics display is suffused with the sounds of wind and water and the movements of rock and snow. The soundtrack of the Views of the Planet video screen features, at key moments, the drip of water and other Earth sounds synchronized with the video. In the area exhibiting black smokers, ores, and different types of weather, all the sound is projected from below. In the black smoker section, there is no sound within the range of human hearing; vibrations projected from a subwoofer under the platform create an atmospheric pressure that gives the impression of being underwater. The exhibits on climate change and weather have an undersea soundscape with surface winds. Visitors hear only a trickling stream and the geologist's pick where rock formations and ores are displayed.

Near the center of the Hall, at the Dynamic Earth Globe, an abstract, lyrical soundscape provides a meditative background to the constant changes on the globe. Speakers set in the ceiling project sound layers emblematic of the forces at work—the sound of wind as Earth's cloud cover is stripped away, the sound of ocean water as the seas drain, and a vibrating gong as the fundamental Earth is laid bare. Voice labels on a separate playback describe the activity shown on the globe; each of the phrases is triggered by cues imbedded in the software that runs the exhibit.

In addition, there is also a twenty-minute ambient soundscape that creates an ever-changing tapestry of Earth sounds—winds, hail, rain, flowing waters, volcanic and gas explosions, lava flows, rock slides, earthquakes, avalanches, mudslides, and calving glaciers. Since Earth's natural sounds are in motion, this soundscape is kinetic, delivered by multiple speakers placed around, below, and above the visitors.

In certain areas, the designers sought only silence, so the soundscapes include strategic pauses. And to limit the total sound in the Hall, each of the video monitors—all of which are equipped with speakers on or immediately above them—has sensors that switch on the audio only when a visitor is within range.

Charles Morrow Associates produced all the soundscapes. The installation was coordinated by Frank Rasor, media manager of the Museum's Exhibition staff, with consultant Brad Berlin of Berlin Acoustics. Electrosonic's Oliver Pemberton provided control system design, installation, and programming.

Visitors relax under the Dynamic Earth Globe as they listen to the lyrical soundscape.

A DIGITAL SPACE SIMULATION

The new Hayden Planetarium is home to the 429-seat Space Theater, the world's largest, most powerful, and highest-resolution virtual-reality facility, similar to those used by the U.S. military to train pilots and astronauts. But this theater takes viewers to corners of space the military won't visit for some time to come, if ever.

Inside the theater is the Digital Dome, a 68-foot-wide curvilinear screen. Images of the astronomical objects displayed on it are stored on a Silicon Graphics Infinite Reality Onyx-2 supercomputer, one of the largest supercomputers dedicated to visual simulations. The supercomputer's twenty-eight central processing units can simultaneously process fourteen gigabytes of data—the equivalent of about two hundred desktop computers—and its storage systems hold up to two terabytes of information.

The Space Theater's regular half-hour shows are streamed from the Onyx-2 via seven fiber-optic channels to seven computer graphics projectors, the largest cathode-ray tube projectors available. Five projectors are positioned 72 degrees apart around the periphery of the Dome; the other two point up. Together, they provide full coverage of the Dome.

The Digital Galaxy, a database of the universe beyond our night sky—including a detailed virtual reality model of the Milky Way—is stored on the Onyx-2. This computer is used not only in the public shows, but also as a research tool for staff astrophysicists, who log on to conduct simulations of, for example, how stars form, how they shape the interstellar gas in our galaxy and in others, and what happens when asteroids or comets strike planets. Few astrophysicists anywhere have an entire supercomputer at their disposal. The Digital Galaxy was created by the Museum's science visualization team and simulation software engineers at Aechelon Technology, Inc., and was funded by the National Aeronautics and Space Administration.

The Digital Galaxy also incorporates luminosity and location information for all the brighter stars within 1,000 light-years of Earth—some 100,000 stars. But since only a tiny fraction of the 100 billion stars in the Milky Way, to say nothing of other galaxies, has ever been observed from Earth, Museum astrophysicists also undertook a massive two-year initiative to statistically model parts of the galaxy that have never been observed directly. That simulation shows the area within 100,000 light-years of Earth: about a third of the galaxy's pulsars, half of the gaseous nebulas, and 80 percent of the stars, about two billion of them. The Digital Galaxy also models about 3,000 galaxies within 50 million light-years of Earth.

The scale of the objects modeled by the Digital Galaxy project is difficult to imagine. If our solar system—the area enclosed by the orbit of Neptune—were the size of a penny, says Jim Sweitzer, director of special projects for the Planetarium, then the Milky Way galaxy would span the distance from New York to Denver.

Museum staff built the digital collection of stars, galaxies, nebulas, black holes, and interstellar gas clouds from a number of sources, including the National Aeronautics and Space Administration, which houses data from many researchers. Most of the information about nearby stars (those within 1,000 light-years of Earth) came from the High Precision Parallax Collecting Satellite, known as Hipparcos, launched in 1989 by the European Space Agency, and named for Hipparchus of Nicea, a second-century B.C. Greek mathematician who compiled the first catalogue of stars and planets.

While the Milky Way is the principal focus of the Digital Galaxy, it is only one of some 100 billion known galaxies. Galaxies are now known to form a network of filaments, a pattern called the large-scale structure of the universe. R. Brent Tully, professor of astronomy at the University of Hawaii, contributed a database of some 30,000 galaxies that illustrate this filamentary pattern in the local supercluster of galaxies. Beyond the local supercluster, Museum staff relied on computer simulations and theorizing by Jeremiah Ostriker, provost and professor of astronomy at Princeton University. At this level, astronomers are concerned with whether the universe is expanding, contracting, or standing still, and Dr. Ostriker coined the term "dark matter" to describe the intergalactic stuff that may be slowing the universe's expansion.

The Hayden Planetarium is the first planetarium in the world to access scientific data sets directly and turn them into meaningful images accessible to the public. But that astronomical information can be thought of as only one of a number of possible data sets that could be displayed in the Space Theater. Live telemetry feeds from NASA satellites or explorers could be shown on the dome, as could images from high-definition television, on-line news of current scientific events, Internet feeds, or any of a number of other possibilities. When information is of broad general interest to the public, as in the case of the Mars Pathfinder landing in July 1997, live feeds will be projected onto the Digital Dome.

The Digital Galaxy will remain a work in progress for years to come. As astronomical science generates reliable information about increasingly distant corners of the universe, Museum astrophysicists will replace statistically generated information with real observations.

The Infinite Reality Onyx-2 supercomputer.

A UNIQUE INSTRUMENT: THE ZEISS PROJECTOR

The Hayden Planetarium also features a one-of-a-kind star projector—the Zeiss Mark IX Hayden Edition—built to the Museum's specifications by Carl Zeiss Jena, the famed German optical works. The Museum had originally ordered a Mark VIII projector, but it requested so many enhancements that Zeiss created a new model number for this instrument. Thirty motors controlled by forty-five computers move and angle the four-ton Zeiss, which took two years to design and arrived in New York in fourteen crates.

It is the latest of four Zeiss planetarium projectors that have been housed at the Museum and occupies the same site as the first (a Mark II installed in 1935), which "looked like a big bug," in the childhood memory of Neil de Grasse Tyson, director of the Hayden Planetarium. Fine holes drilled in copper plates produced the stars seen on the old Planetarium's green ceiling, and a technician moved the tracking system with a single motor. The new Mark IX projects high-intensity white light through fiber optics onto about thirty fixed-image plates in the Digital Dome and shows the night sky, some 9,100 stars, as well as other celestial objects, more clearly, sharply, and realistically than any other projector. It features a markedly improved view of the Milky Way as seen from Earth and the first-ever astronomical presentation of all planets of the solar system, including Uranus and Neptune, which are not visible to the naked eye, as well as star fields, the sun, and its planets from the surface of any object in the solar system. Among the other unique features of the starlit sky shown by the Hayden projector are deep sky objects—faint galaxies, star clusters, and gaseous nebulas. Star maps are based on data from NASA. A special projector demonstrates the supernova that unexpectedly lit up in the Southern Hemisphere in 1987. Standard Zeiss drawings of the constellations were replaced by new drawings rendered by Scott Ewalt, a New York artist. Using microelectronics, Zeiss was able to place all the instrument's effects into one projector-studded "star ball" instead of the two in older versions.

The stars of the new projector are equipped with "scintillation," a randomized dimming device that recreates the twinkling of starlight shimmering in Earth's turbulent atmosphere. The instrument can also demonstrate precession, the wobble in Earth's axis of rotation, which changes its orientation to the sun over thousands of years.

Because light pollution and haze obscure the night sky in New York, Neil de Grasse Tyson says, "It is particularly important that city residents have an accurate Planetarium sky."

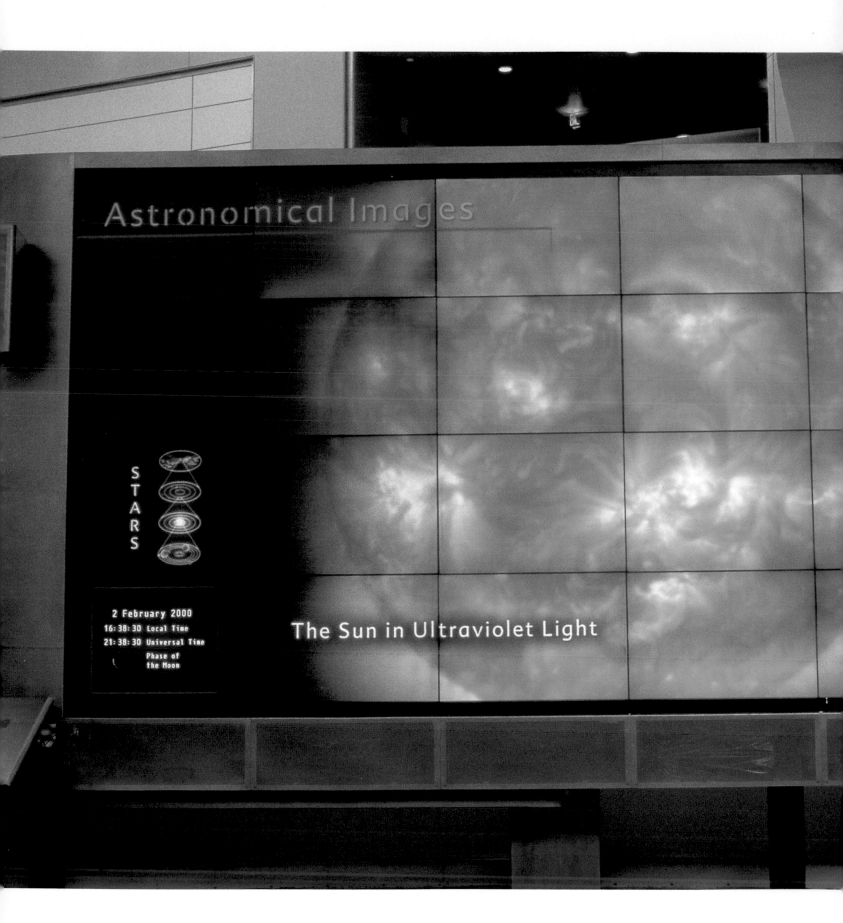

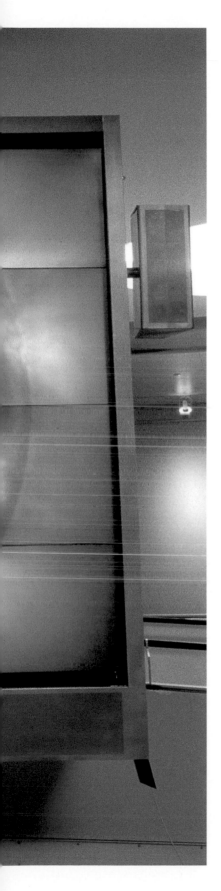

OFF-THE-WALLS VIDEO

The Rose Center benefits from the groundbreaking work of the Science Bulletins department. Every day, its staff produces short video pieces that resemble television news stories. But these pieces are shot in high-definition format and can be viewed only at the Museum's three video walls—two of which are in the Rose Center—interactive video kiosks, and on a special website, http://sciencebulletins.amnh.org. The Museum is one of the few science institutions anywhere to produce daily news features on scientific topics.

The Science Bulletins team relies on Museum scientists and live video feeds from ABC News, NASA, and the Weather Channel to produce its daily updates on Earth science, astronomy, and biodiversity. For the Earth Event Wall in the Hall of Planet Earth and the Astro Event Wall in the Hall of the Universe, a graphics team uses Everest software, a real-time, three-dimensional high-resolution authoring tool, on the Museum's Onyx-2 supercomputer to create high-definition maps and computer graphics that demonstrate the science behind the news.

The Science Bulletins unit can send video production crews anywhere in the world at a moment's notice. Images are shot on high-definition videotape, edited as digital video, and then compressed as MPEG-2 computer video for internal use or transfer. A digital disk recorder, essentially a computer video server, decodes the MPEG-2 video

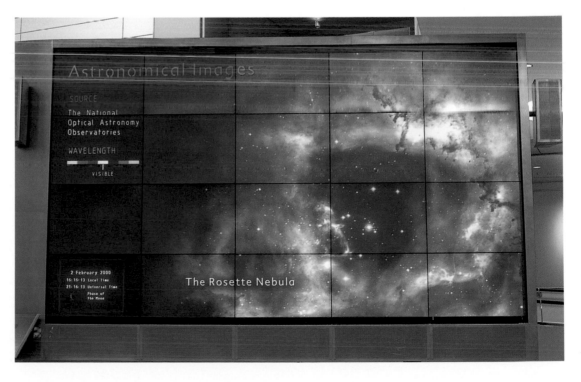

sequences and supplies them to high-definition projectors in each exhibit hall. Within hours of an event—a meteor impact in Alaska, for example—news about it can be displayed on one of the video walls in brilliant high-definition images. The website and kiosks in the Museum, which also archive the stories, take their feeds from the same source but show the material in standard definition.

The Earth Event Wall in the Hall of Planet Earth, which measures 7.5 by 11.5 feet, offers news of volcanoes, earthquakes, and other geologic events. A news segment on an earthquake in Turkey, for example, used video clips, graphics, and text to pinpoint the location of the Anatolian Fault Zone, the most active transform fault in the world, and to demonstrate how the movement of tectonic plates causes earthquake activity in that region.

The Astro Event Wall in the Hall of the Universe, at 9 by 13.5 feet, the largest of the three walls, displays an eight- to ten-minute loop of news stories about astronomy and the universe on such topics as colliding galaxies, extrasolar planets, and volcanic activity on Jupiter's moon Io. That video wall consists of twenty rear-projection cubes controlled by an Electrosonic matrix box.

One important source of information for the Rose Center will be NASA's new Earth Observing System, an array of satellites that promise to offer scientists important data on climate change that may be taking place as a result of human activities. The first such satellite, Terra, launched in December 1999, will study clouds, water vapor, aerosol particles, trace gases, land surfaces, and oceans. Terra and its sisters will collect information such as temperature via remote sensors and relay it to Earth, where someday soon it may become the topic of one of the Museum's Science Bulletins.

Smokey Forester, Science Bulletins manager for the Museum, notes that television news can devote only a limited amount of time to science topics. The Museum's role is to examine science stories in greater depth, often navigating through reams of data, interpreting them and offering graphic representations that capture the imagination. Students are given the opportunity to draw connections between what they read in textbooks and current science. Educators can use the materials to spark interest in science.

The Science Bulletins project is part of the Museum's National Center for Science, Literacy, Education and Technology, founded in 1998 with funding from NASA. The center produces curriculum material, offers the science bulletins on the web, and plans to send its up-to-the-minute video productions over DSL lines to smaller museums around the country.

EXPLOSIVE ORIGINS

Scientific understanding of the processes that brought the universe into being continues to evolve. In a brief laser show of dazzling intensity, the Big Bang Theater in the Rose Center presents the latest views of this seminal event, now generally acknowledged to have taken place approximately 13 billion years ago.

Until the twentieth century scientists resisted the idea that the universe had a beginning. But the work of Albert Einstein, the American astronomer Edwin Hubble, and the Austrian Christian Doppler indicated that the universe was expanding. In 1946, George Gamow, a Russian-born scientist, proposed that this movement was the result of a primeval fireball, an intense concentration of pure energy that was the source of all matter and dimension. In 1965, Arno Penzias and Robert Wilson of Bell Labs encountered a noise in the universe that turned out to be the cosmic background radiation, an infinitesimal amount of energy that permeates the universe and is the fossil remnant of the Big Bang. A volume of space smaller than a grain of sand has expanded to become our entire observable universe. How? British cosmologist Stephen Hawking has written that "the actual point of creation lies outside the scope of presently known laws of physics." We simply don't yet know.

At the Big Bang Theater, in the bottom half of the Hayden Sphere, viewers witness a recreation of the birth of the universe. Standing on a Plexiglas floor that covers the central 36-foot-wide screen, they watch the frenzied origin, expansion, and cooling of the universe brought to life through a brilliant laser display, dozens of lighting effects, an LED display, and an explosive soundtrack including narration and surround sound to immerse visitors in the imagery and energy of the early universe.

Laser projectors produce a unique combination of color integrity and beam sharpness that can create the illusion that there is no screen surface, that images are floating in air. Because the laser beam produces very little light scatter where it meets the screen, and thus no ambient light, viewers have the illusion that they are looking out on a bottomless window into the early universe.

The laser used in the theater's Omniscan projector is a Class IV continuous wave multiline design, equivalent to a 16,000-watt lightbulb, selected and installed by Singularity Arts. The multiline instrument produces several different colors of light, which can be mixed using a high-frequency oscillating crystal to create any color. Computers control the laser, directing two scanning mirrors to "paint" the laser's light onto the screen, in much the way a moving flashlight beam seems to create a circle on the wall. Very intricate imagery that appears to be coming from multiple sources can be created through "blanking." The scanning mirrors paint one section of the screen, then jump to another while the beam is blanked, or shut off, for a fraction of a second. It happens so quickly that the eye does not detect it.

Visitors leaving the theater can continue on an awe-inspiring journey that chronicles the evolution of the universe by following the Cosmic Pathway, the sloping walkway that takes you through 13 billion years of cosmic evolution.

Much of the data in the Rose Center is available to the general public through the Museum's website, http://www.amnh.org, or at the Museum's Science Bulletins site, http://sciencebulletins.amnh.org.

Lights out in the Big Bang Theater.

PHOTOGRAPH CREDITS

All photographs were taken by Denis Finnin, director of the American Museum of Natural History photo studio, except as noted below:

The Judy and Josh Weston Pavilion opened in February 2001. It extends the Rose Center facility westward and provides the Museum's first public entrance on its west side. The centerpiece of the Pavilion is an 18-foot, 3,500-pound sculpture of an armillary sphere, designed by the Museum's Exhibition Department. Early armillary spheres, dating back to the 2nd century A.D., were used to derive the coordinates of stars and planets. Later models served as geocentric teaching aids, demonstrating the motion of celestial bodies as viewed from Earth. This aluminum and stainless steel sphere, which places our Milky Way galaxy at the center, is positioned to demonstrate New York City's precise location on January 1, 2000.